RANCH SHIPMENTS

D1706341

```
                                    EKA
j599.55
GREENE      Greene, Carol.
                Reading about the
              manatee

            19.95 ; 5/96
```

HUMBOLDT COUNTY LIBRARY.

DEMCO

# Reading About
# THE MANATEE

## Carol Greene

**Content Consultant:**
Dan Wharton, Ph.D., Curator,
New York Zoological Society

**Reading Consultant:**
Michael P. French, Ph.D.,
Bowling Green State University

ENSLOW PUBLISHERS, INC.
Bloy St. & Ramsey Ave.　　P.O. Box 38
Box 777　　　　　　　　　　Aldershot
Hillside, N.J. 07205　　　　Hants GU12 6BP
U.S.A.　　　　　　　　　　　　U.K.

Copyright © 1993 by Enslow Publishers, Inc.

All rights reserved.

No part of this book may be reproduced by any means without the written permission of the publisher.

**Library of Congress Cataloging-in-Publication Data**
Greene, Carol.
    Reading about the manatee / Carol Greene.
      p. cm. — (Friends in danger series)
    Includes index.
    Summary: Describes the physical characteristics and behavior of the Florida manatees, or sea cows, and discusses some of the dangers they face.
    ISBN 0-89490-424-8
      1. Trichechus manatus—Juvenile literature 2. Endangered species —Juvenile literature. [1. Manatees. 2. Rare animals.]
    I. Title. II. Series: Greene, Carol. Friends in danger series.
    QL737.S63G74 1993
    599.5'5—dc20
                                                                                        92-26811
                                                                                                    CIP
                                                                                                     AC

Printed in the United States of America

10 9 8 7 6 5 4 3 2 1

**Photo Credits:** ©D. Holden Bailey/Tom Stack & Associates, pp. 1, 8; ©Douglas Faulkner/Photo Researchers, Inc., pp. 4, 6, 14, 16, 24; ©Jeff Foott/Tom Stack & Associates, p. 10; ©Francois Gohier/Photo Researchers, Inc., p. 20; ©Richard Hutchings/Photo Researchers, Inc., p. 26; ©Andrew J. Martinez/Photo Researchers, Inc., p. 12; ©Patrick M. Rose/Florida Department of Natural Resources, pp. 18, 22.

**Cover Photo Credit:** ©Andrew J. Martinez/Photo Researchers, Inc.

**Photo Researcher:** Grace How

# CONTENTS

Happy . . . . . . . . . . . . . . . . . 5

Danger! . . . . . . . . . . . . . . . 21

What You Can Do . . . . . . . . 27

More Facts About
the Manatee . . . . . . . . . . . . 29

Words to Learn . . . . . . . . . . 31

Index . . . . . . . . . . . . . . . . . 32

# HAPPY

They look like clouds,
big gray clouds,
sleeping on the bottom
of a Florida river
near the Atlantic Ocean.
They are manatees.

Slowly, two manatees
rise to the surface.
(Manatees do everything slowly.)
They breathe, then slowly
sink to the bottom again.
They don't even wake up.

A large group of manatees rest on the bottom of a river in Florida.

These two manatees are Happy
and her baby, or calf.
Happy is ten feet long.
She weighs 1,000 pounds.

Her calf is three feet long.
He weighs 60 pounds.

Their huge size is one reason
manatees move so slowly.

Happy and her calf
will sleep all day
with others in their family.
Then, when night comes,
they'll wake up to eat.

A manatee and her calf.

Slowly, slowly,
they'll swim along.
Happy's rounded tail
pushes her through the water.
She uses her short front
flippers to turn herself.

The calf swims only
with his flippers.
He doesn't know how
to use his tail yet.
Sometimes he rides
on his mother's back.

Manatees use their rounded tails to swim.

Happy has nails on
the end of her flippers.
They help her find
a clump of water plants.

She grabs some plants
with her upper lip
and puts them in her mouth.

Happy's upper lip
is split in two.
She uses it like a clamp.
It's almost like
an elephant's trunk.

Manatees and elephants
share an ancestor.
Elephants are the manatees'
closest relatives.

A manatee chews on a plant.

Happy will eat 100 pounds
or more of plants tonight.
Those plants would clog
the river if manatees
didn't eat them.

The calf eats plants too.
But he also drinks
his mother's milk.
Manatees are mammals.
The mother's body makes
milk to feed her baby.

A manatee nurses its calf.

Happy and her calf
swim past a strange heap
on the bottom of the river.
Happy goes over
to see what it is.
Manatees are curious.

The heap is just trash.
Happy leaves it.

A curious manatee searches the bottom
of a river.

Next, she meets
another manatee.
They bump noses.
That is how manatees
say hello to one another.
They are friendly animals.

On they swim.
Happy and her calf
will swim and eat
until morning comes.

Two manatees greet each other.

Now and then,
Happy makes a noise.
Some people say
it sounds like a groan.
But others know it is a song—
a slow manatee song.

No one is sure why manatees make their noise.

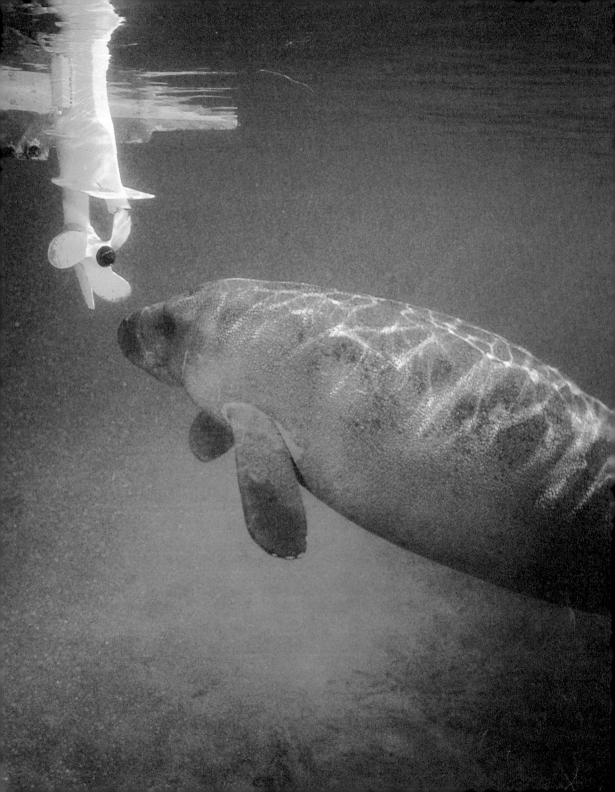

# DANGER!

Manatees do not hurt
any other animal.
No other animal
hurts them—except people.

People drive barges
and fast boats through
the Florida waters.
They don't see the manatees,
so they hit them.

Some manatees die.
Others are hurt.

A manatee swims too close to a dangerous boat.

There are about 1,200
manatees left.
Scientists have pictures
of 900 of them.
They all have scars
from boats and barges.

That's a lot of hurt manatees.

Trash in the water
can hurt manatees too.
They can swallow
plastic bags and fish hooks.
These hurt their insides.

This manatee has many scars from boats.

Wastes dumped into the water
can also hurt them.

But manatees' worst enemies
are those boats and barges.
If the people who drive them
are not more careful,
soon there will be
no manatees left.

But if they leave
them alone,
the manatees will do fine.

Scientists are studying the few manatees left.

# WHAT YOU CAN DO

1. Learn more about manatees. Read books and watch nature shows.

2. Don't dump trash in the water. Put all trash where it belongs. Even if you do not live near any manatees, your trash might hurt other kinds of animals.

3. See if your family can join a group that works for manatees and clean water. Your librarian can help you find names of these groups.

One way you can help the manatee is by keeping trash out of the water.

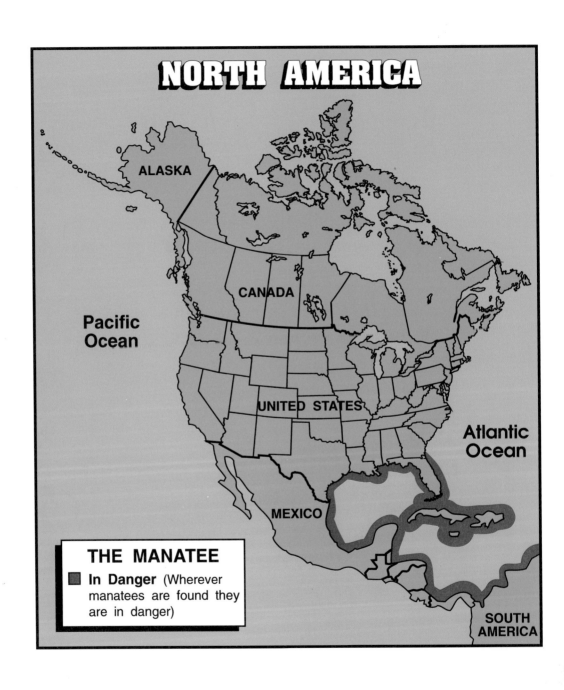

# MORE FACTS ABOUT THE MANATEE

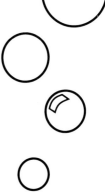

- In the United States manatees live as far north as the coastal waters of North Carolina. They live as far south as the Florida Keys. But most stay right in Florida.

- Their relatives, West Indian manatees, move from the West Indies to northeastern South America.

- Florida manatees are also related to the Amazon manatee and the West African manatee.

- Some people call manatees "sea cows."

- A manatee's skin is gray when it's wet. But when it is dry, it's almost white. People long ago who thought they saw mermaids might have been looking at manatees.

# WORDS TO LEARN

**calf**—A baby manatee.

**flippers**—Short, arm-like parts toward the front of the manatee's body.

**manatee**—A large mammal that lives mostly in the waters near the coast of Florida. Its Latin name is *Trichechus manatus latirostris*.

**mammal**—An animal that feeds its young with milk made by the mother's body; it has some hair or fur.

**mermaid**—A make-believe creature that is supposed to be half woman and half fish.

**scar**—A mark left on the skin after a wound has healed.

**sea cow**—A name some people use for manatees.

# INDEX

Amazon manatee, 30
ancestors, 11
boats and barges, 21, 23, 25
calf, 7
    food for, 13
    length of, 7
    swimming and, 9
    weight of, 7
flippers, 9, 11
Florida, 5, 21, 29
mammal, 13
manatee
    appearance of, 5
    breathing and, 5
    coloring of, 30
    eating and, 11
    enemies of, 25
    food for, 13
    greetings of, 17
    length of, 7
    locations of, 29
    numbers of, 23
    people and, 21
    relatives of, 29, 30
    sleeping and, 5, 7
    sounds of, 19
    speed of, 5
    swimming and, 9
    things to do for, 27
    trash and, 23
    upper lip of, 11
    weight of, 7
mermaids, 30
sea cow, 29
upper lip, 11
water pollution, 25
West African manatee, 29
West Indian manatee, 29